CAREERS IN

ROBOTICS

ENGINEERS – TECHNICIANS

RESEARCH

MANUFACTURING

CHOOSING A CAREER SHOULD BE interesting, challenging and even fun. You have more options available to you than almost anybody else in history. Wading through the information may seem like a formidable task, but you should look at the process as one with great rewards at the other end.

Most people tend to evaluate potential careers based upon what they could do right now. Not all careers fit this mold, however. Some are attractive for what they might lead to in the future. Robotics certainly fits into this latter category. Every robot ever built has led directly to a better robot, and sometimes very quickly. Robots now handle most of manufacturing in the industrialized world. Robots even build other robots, which is either a tantalizing thought or a scary one, depending upon how you look at it. And everybody knows that the ultimate goal is to create a robot that is indistinguishable from a human being.

The robotics industry needs capable, imaginative professionals ready to build upon the achievements of the past and create the robots of the future. Robotics engineers are serious dreamers who look at challenges as nothing but tiny speed bumps on the way to the next big breakthrough. Their successes not only move their industry forward, but propel innovation in other industries. Developments in robotics have taken off in recent years, helped along by strides made in software, electronics and materials technologies.

WHAT YOU CAN DO NOW

YOU CAN GET GOING ON YOUR career in robotics right now. The vast majority of the world's robots are used in industry. They are essentially arms attached to computers and would not fit most people's definition of what constitutes a robot. There is nothing humanoid about them, and they are programmed to do only one thing at a time. They don't even look very impressive, at least to the untrained eye. What they do, however, is very impressive. Make an effort to check them out in action. Contact a local factory that uses robots and ask if you can take a tour. Ask a teacher – probably in mathematics, science or shop classes – to arrange a field trip for you and your classmates. See some actual robots up close and then start thinking for yourself. How could you do it better?

In simplest terms, robots are mechanical agents that perform tasks on behalf of people. A thermostat that automatically maintains a steady temperature in your home qualifies as a robot, although an extremely simple one. Ever seen a Roomba robotic vacuum cleaner? Check one out on iRobot.com. Many companies offer robots in kit form for use as science projects. Search the Internet for robot kit, and start building your own robot.

There is no faster, better way to learn about any career than by reading trade journals devoted to it. There is at least one magazine, entitled Robot, devoted entirely to robotics, and several, like Popular Science, Popular Mechanics and Military Technology, that cover the subject regularly.

Be sure to check out movies that feature robots, such as *Metropolis, 2001: A Space Odyssey* and *Forbidden Planet,* among many others.

HISTORY OF THE CAREER

INANIMATE OBJECTS THAT CAN THINK for themselves and be of service have been a part of human mythology for thousands of years. The Greek god Vulcan, for example, created mechanical servants both in human form and in purely practical variations like tables that could move about on their own. Chinese legend speaks of a mechanical human presented to King Mou in about 1000 AD.

Several names have been used to describe robots. Most

robots are more accurately known as automatons, or devices able to carry out specific tasks without human intervention. Automatons can look like anything, including humans, although the name is usually reserved for robots without human traits or humanoid robots with very minimal intelligence. Intelligent robots that look like humans are known as androids.

The word "robot" was first used publicly in 1920 by Czech playwright Karel Capek in his play *Rossum's Universal Robots,* a story about a factory that produces humanoid robots to serve human masters. Modern usage has also coined the term "artificial intelligence" or AI to describe machines that can think for themselves, whether or not they perform mechanical functions.

Generally understood definitions of what does and does not constitute a robot have evolved over time. The earliest robots, for example, would today be defined as mechanical toys, like wind-up mannequins and basic devices like clocks. The key distinction was that they all performed simple functions entirely on their own, without human control. No less than Italian visionary Leonardo da Vinci designed a mechanical knight sometime around 1500, although it is not known if he ever tried to build one.

The first automated machines to be put into widespread use were mechanical looms for the textile industry. By the mid-1800s, most cloth was woven mechanically, resulting in dramatically increased production and reduced prices. Mechanical looms illuminated the best use for automatons – to do dangerous, repetitive jobs that don't require humans.

Arguably the next major step in the development of robotics was fictional. German director Fritz Lang's 1927 silent film *Metropolis* featured the *Maschinenmensch*, a female humanoid robot programmed to wreak havoc. This was the first cinematic robot to capture the imagination of a wide audience, leading to several other movie robots that have stood the test of time. Klaatu in 1951's *The Day the Earth Stood Still,* Robby the Robot from the 1956 film *Forbidden*

Planet, R2-D2 and C-3PO from the *Star Wars* movies, and the Terminator from the movies of the same name.

By the 1950s, the development of robots took place side-by-side with the development of computers. The first industrial robot was installed in a General Motors plant in 1961. Known as Unimate, the robot was essentially a mechanical arm that could perform repetitive tasks indefinitely and with great precision. Unimates could also reach into places humans could not access and work in dangerous environments like paint booths filled with toxic fumes. The mass employment of Unimates and their successors has revolutionized many industries by speeding up production and lowering costs.

Unimates and other early industrial robots were controlled by room-sized computers that could not possibly be installed in the robots themselves. The introduction of the first microprocessor in 1971 paved the way for dramatic increases in computing power and in robot design. No larger than a fingertip, early microprocessors were more powerful than the enormous computers they replaced. This allowed new generations of robots to carry their computers around with them.

Modern industrial robots today perform almost all of the assembly of complex mechanical products, such as automobiles. They are accurate, able to work very long hours in environments that would be dangerous to humans and, most importantly, cost much less over time than human workers. Employment of robots has dramatically reduced the cost of building automobiles and many other products. That's why there are about one million industrial robots in use worldwide today, with about half of them in Japan.

While industrial robots account for most of robots in use around the world, many other robots have been created to serve a wide variety of purposes. All of the spacecraft launched by the National Aeronautics and Space Administration (NASA) since the 1970s have had some robotic function, from Voyagers I and II launched in 1977 to independently probe our solar system and beyond, up to the Mars rover Curiosity launched in 2011 to perform complex scientific experiments on the Martian surface.

Robots are routinely used by the military and police to do jobs too dangerous to be done by humans. In the military, airborne drones are used to conduct automated surveillance over combat zones and even to drop bombs without putting a human pilot in harm's way. The United States military is also developing four-legged robots to serve as mechanical donkeys capable of carrying heavy loads over irregular terrain.

Police use small robots on treads, for example, to search areas where they think armed criminals are hiding. The robots, which are equipped with cameras and other sensors, can tell officers where suspects are hiding without fear of being shot.

Some robots are built just to prove that they can be. The IBM computer Deep Blue, for example, was constructed for the single purpose of beating human chess champion Gary Kasparov in a game of chess. In 1997, on the second try, it did. In 2000 Honda introduced ASIMO (Advanced Step in Innovative Mobility), a humanoid robot and technology demonstrator that the company hopes will lead to a practical, affordable robot that can be a mechanical helper to people with disabilities. In 2002 a company called iRobot introduced the Roomba, a robotic vacuum cleaner that can vacuum an entire house without human intervention. In 2011 NASA launched the Robonaut 2 to the International Space Station to work alongside humans in maintaining the space station and conducting scientific experiments.

With the advent of ever more powerful microprocessors and new materials technologies, the field of robotics is entering a fascinating new stage. Industrial robots have become a mature technology able to grow and improve steadily, while humanoid robots are on the cusp of practical reality. It is almost certain that complex humanoid robots will become a reality during your working lifetime. You can be a part of making the dream come true.

WHERE YOU WILL WORK

THERE IS NO REAL GEOGRAPHIC CENTER of the robotics industry, although there are large concentrations of employers in the Northeast and Midwest. Many robotics companies are located in the state of Michigan. Detroit has long been home to the American automobile industry, which also happens to be the single-largest user of industrial robots and other forms of automation. Not all automobile manufacturing plants are located in Michigan but it's the home of the Big Three – General Motors, Ford and Chrysler.

Their own creations have allowed artificial intelligence software companies to become almost virtual, with designers writing code from home offices all over the world. Serious research and development of AI applications happens at many universities and in many businesses across the country. There is a dense concentration of companies in California's Silicon Valley region near San Francisco. Not all of these companies are directly involved in robotics but artificial intelligence is a critical component of robotic engineering, especially at the highest levels.

YOUR WORK DUTIES

ROBOTICS IS A VERY COLLABORATIVE field in which professionals from many disciplines work together toward a common goal. A person with the title of "robotics engineer" may actually have a degree in electrical engineering, mechanical engineering or some variation on computer science or artificial intelligence. As you look through these job descriptions try to think of robotics as a career you grow into rather than a career you train for specifically.

Mechanical Engineer

Mechanical engineering is a very broad field that covers many engineering applications. Mechanical engineers design and build machines of all sorts, from internal combustion engines to machine tools, to complex factory equipment and robots.

Mechanical engineers are responsible for the mechanics of a given project. In the case of robotics, that could mean designing and building a complex armature system for an industrial robot to be used in a custom application on a factory floor, in a spacecraft or in an experimental humanoid robot. A mechanical engineer may create an apparatus for an industrial robot that can grip a very specific type of auto part and install it perfectly on an assembly line thousands of times each day. A more complicated variation may require a mechanical engineer to design an apparatus that can grip several different kinds of parts to be installed in different kinds of cars rolling through a common assembly line.

Mechanical engineers add real utility to robots by creating the limbs and other tools that enable them to do things. Their efforts are often the most visible when finished products are rolled out to customers or the general public. Speech software has become so common that nobody is impressed that Honda's ASIMO robot can speak, but everybody notices that it can walk with a very human-looking spring in its step. This is what mechanical engineers bring to the project.

Electrical Engineer

Mechanical engineers wouldn't get very far without the electrical engineers who figure out how to bring power to their creations and bring them to life. Electrical engineers devise the controls, motors and power supplies that provide energy and motive force to robots.

For example, an electrical engineer may be assigned to work

with the mechanical engineer on the gripper apparatus for an industrial robot. The electrical engineer will figure out how to get enough power and motive force into the apparatus in order for it to do its job. This includes making sure those necessary items like motors and controllers fit within the space available on the apparatus. Not all electrical items have to be located on or within the apparatus, but some do and they have to help achieve the end goal, not hinder it. A gripper that requires an enormous amount of power at its fingertips, for example, may pose a challenge for both the mechanical and electrical engineers as they try to figure out how to mount motors with enough power in an apparatus that has to fit into a relatively small space in an automobile rolling down an assembly line.

Electronics Engineer

Electronics engineers are closely related to electrical engineers but tend to specialize in control systems and other complex electronics, like global positioning systems and communications devices. Robots may use complex electronics systems, depending upon their function. The assembly line robot may be equipped with a sophisticated visual reference system that allows it to make minute adjustments to the parts it installs. Much like a human eye, such a system enables the robot to see what it is doing, just like humans would use their eyes to help them install a part. Electronics engineers generally develop a specialty during their careers.

Computer Hardware Engineer

Computer hardware engineers concentrate their efforts on hardware specifically for computer systems. All robots require some computer hardware to function. Some robots carry their computer hardware with them. Humanoid robots, for example, must carry all of their parts with them in order to function independently. Most industrial robots are connected to computers that may be in completely different rooms. For a computer hardware engineer specializing in robotics, the ultimate challenge is to design and build hardware that fits into mobile robots.

Computer Software Engineer

Software engineers specializing in robotics provide their creations with the brains to do their jobs. They write the applications that enable robots to perform simple or sophisticated tasks over and over again, exactly the same way, without fail, year in and year out.

This is more complex than it may seem. If the assembly line robot needs to be able to grip several different kinds of parts and install them in several different types of cars and requires a visual electronic device to do it, the software that enables all of those functions will be extremely complicated. Each function will have to be integrated with all the rest. The electronic eye will have to be able to recognize the parts for each different type of car and the specific points at which they are to be installed. The entire system will have to change every few minutes as a different car rolls down the assembly line. Such software would also require a reporting mechanism to keep track of parts installed, and a diagnostic system to prevent breakdowns. The software may even keep track of the robot's more mundane functions, like operating temperature and lubrication.

Industrial Designer

Industrial designers play a small role in robotics but it is an important one. Industrial designers design all of the outside casings and pieces of industrial equipment, including robots. Industrial robots don't have to be particularly beautiful but they do have to fit into a busy industrial environment. Industrial design becomes especially important when designing robots that will have some public use, such as technology demonstrators like ASIMO or even robot science kits designed to appeal to teenagers. Robotics is still a young profession in many ways. Capturing the attention of the public with dramatic design can be helpful.

Robotics Technician

There are more robotics technicians than there are robotics engineers. Robotics technicians are responsible for maintaining robots and making sure that they run at peak efficiency.

Most robotics technicians work for industrial employers or for service firms available to repair and maintain robots on an on-call basis. Some robotics engineers double as robotics technicians, especially when working on technology demonstrators or one-of-a-kind robots like those used by NASA. But when fleets of robots are deployed throughout a factory environment, there will be a number of robotics technicians on hand to keep them up and running.

Robotics technicians typically earn bachelor's degrees in robotics technology, or mechanical or electrical engineering, and then pursue a career path as a robotics technician. Many earn certifications in specific types of robotics and carve out very comfortable and rewarding careers dealing with a certain type of robot and customer.

Robotics Entrepreneur

Because the robotics industry is always looking for the next big breakthrough, it is entirely possible to make your own way in the business. You will almost certainly have to work for others for a few years and get to know the business and the details of your specialty. But if you have a good idea and can get the financing necessary to get it into production, there is no lack of opportunity within the robotics business for inventors and innovators who want to push the envelope.

ROBOTICS PROS TELL THEIR OWN STORIES

I Am a Robotics Engineer for NASA

"I have one of the coolest jobs in the world. Seriously, I sometimes look at the stuff I do and wonder if I'm watching a sci-fi movie. Then I remember that what I'm looking at is the real life that I get to live every day.

We are really at the cutting edge here at NASA. We attract the best and brightest from the nation's best universities and put them to work solving the problems posed by space exploration. This is a wonderful combination.

I started to head in this direction when I was in high school. I always did well in math and science and loved science fiction. Don't discount the value of sci-fi in the development of people in professions like mine. Good sci-fi is nothing more than a look at what the world may be like in the future when technology has progressed far beyond where it is now. Technology will solve some problems even while it creates others. Science fiction gives people like me something to aim for. Remember the communicators on the original *Star Trek* television series, and how they flipped open? Remember the first pocket-sized cell phones? They flipped open too. That wasn't a coincidence.

I majored in mechanical engineering in college and started out with a small industrial robotics company. I learned a lot about the basic mechanical principles behind robotics and got a good picture of what was possible and what was difficult. I won't admit that anything is impossible. I started graduate school after working for a few years and set my sights on a career with NASA. It seemed like a longshot, but after earning a Master of Science degree and PhD and proving myself with several years in industrial robotics, I managed to get my foot in the door.

What a wild ride it has been. I have been on teams that designed armatures for the space shuttle fleet, automated landing gear for rovers launched toward other planets, and even delicate testing and sensing devices for rovers charged with performing scientific experiments after landing. The breadth of my experience here has been incredible.

For example, the armature for the space shuttle has to be strong enough to manhandle huge satellites out of orbit and into the shuttle bay so they can be repaired by mission specialists. There may not be much gravity in space but that's not really the point. The armature has to be able to grab and hold something that weighs several tons and very carefully place it inside the bay without damaging either the satellite or the space shuttle. This is very delicate work, but on a grand scale.

I got to work on designs for a new Mars rover. When it lands on Mars, this rover will roam the planet's surface, stopping once in a while to conduct scientific experiments on the soil and atmosphere. The goal is to determine if there is or ever was life on the Red Planet. The robot consists of an arm that can reach down into the Martian soil like a small shovel, retrieve a sample and bring it back into the rover where it can be carefully parceled out among half a dozen different science experiments. Each experiment consists of its own robot. There will be no human hands to help the rover do its job, you know. Each experiment will subject its portion of soil to a specific type of analysis, register the results, transmit the data back to earth and then clean itself up and prepare for the next sample. A pretty amazing feat!

I'd recommend this career to anybody who is truly fascinated by robots and the possibilities they represent. It's not an easy profession, by any means. Devising robots for space exploration is something very few people can do. Without the sense of wonderment and limitless possibility no one would tackle it."

I Am a Mechanical Engineer for an Industrial Robotics Company

"Most people would not immediately identify our products as robots. It's sort of a running joke in the industrial robotics business that we will never truly satisfy our customers until our products look like humans. There is no practical reason to make industrial robots look like people, but the popular, sci-fi vision of robots is so universal that the idea just won't go away.

The truth is that most industrial robots are little more than very smart arms. They are usually equipped with a specialized tool at one end and a powerful computer at the other. They do things that humans can't or won't do, and they do them 24 hours a day, seven days a week. They cost a lot to buy and install, but once they're up and running they cost much less than human employees and are much more reliable. You can't be late if you never leave the building.

I got into the business after earning a master's degree in mechanical engineering and taking a job with a robotics company. I hadn't really planned on a career in robotics, but this was my first job offer, so I took it. Turned out to be a really good thing, as I've been at it ever since.

Industrial robots are more complicated than they seem. If you watch a modern assembly line you'll see a long line of robots effortlessly spinning through their assigned tasks in what looks like a very precise mechanical ballet. The idea is to keep it as simple as possible. Achieving that, however, is difficult.

Inserting a small precise devise into a monocoque auto body, for example, requires many steps. A monocoque is a type of vehicle in which the body is combined with the chassis as a single unit. One of the advantages that robots have over people on assembly lines is that they can reach into places that humans can't or can't without great difficulty. So a robot can install something inside an auto body through an access hole too small for a human hand. When it gets there it needs to place the item exactly where it needs to be and attach it, either by welding, screwing or pressure-fitting. To do this, the robot and the moving assembly line must be in perfect synchronization and the auto body must be in exactly the same place on the line every time. Because these arrangements are never quite as perfect as we would like them to be, the robot must be equipped with an electronic eye that can see the part and its target location and fine-tunc the assembly as it progresses. This may have to happen more than 1,000 times each day, without a single mistake. My company designs and builds the robots that deliver this kind of precision.

I take great pride in my work. It provides me with enormous intellectual satisfaction, for one thing. It's challenging and often just plain hard, but I always manage to figure it out. Much of the credit has to go to the great team of people around me. This is a very collaborative business. I couldn't possibly do this alone."

I Am a Robotics Engineer Developing Humanoid Robots

"I work for a major university that is working on developing humanoid robots. We are the centerpiece of a large consortium that includes specialists from industry and government, as well as researchers and academics like me. I am one of the permanent team members assigned to the project. Our colleagues from industry and government tend to rotate in and out, bringing good ideas with them every time.

I have a PhD in robotics engineering. I've never wanted to do anything else. I have a very intellectual approach to my field, and set out to work in a university setting after graduating from college. I've been very lucky to maintain this working arrangement for many years now.

Ask just about anybody if humanoid robots – androids – would be a good thing to have and you will get an enthusiastic 'yes.' Ask the same people why androids would be so good to have around and you'll get a lot of blank looks. The idea of having androids to do jobs we don't want to do is so deeply embedded in the human psyche that we can't divorce ourselves from it. Hey, don't get me wrong. I'm in that category, too.

All sorts of potential applications have been dreamed up for androids. Some of them are very utopian and pleasant-sounding, like having android companions for elderly people who live alone and need round-the-clock caregivers. Others are a little sinister, like using androids as soldiers to fight wars.

Designing humanoid robots presents all sorts of challenges. Balancing on two legs requires an inertial guidance system small enough to fit inside a human-sized robot. Arms strong enough to lift anything need to have small motors powerful enough to be useful and also fit inside the human-sized case along with everything else. Since truly independent androids can't be tethered to power cables, androids also need to be able to carry their own power source. And let's not forget eyes, ears and a sense of touch. So far nobody has suggested that androids really need a sense of smell or taste, but they do need to be able to see and hear, and without a sense of touch a powerful android could accidentally crush anything it touches.

Many of the advances we have made in this project have actually been used in other projects. We have been good at figuring out how to get maximum utility into minimum space, for example, and many of the small devices we have created have been modified for use by NASA and for medical purposes. There won't be a market for androids until they become useful enough and cheap enough to appeal to a large market. For now, androids remain very primitive, stupendously expensive and confined mostly to laboratories and theme parks. That's okay for now."

I Am a Robotics Technician for an Automobile Company

"Some people love to hear my story. Some people hate to hear my story. But it's my story and I'm sticking to it.

I spent my youth working on an automobile assembly line in Michigan. Hardcore, blue-collar work, it was. We punched in at the same time every day, did the same repetitive work every day, punched out at the same time every day. Many of the people around me romanticized our existence, trying to make us believe that there was something pure and honest about doing boring work day in and day out.

The biggest problem looming on the horizon, many among us believed, was the coming of robots. 'Robots will take away our jobs,' they said. 'Quality will suffer and our way of life will disappear.' Funny thing was, Japanese auto companies installed robots a long time before we did. Their cars were better and their people, for the most part, learned to do other things.

Eventually the robots did come. At first they were used only for jobs that were dangerous for people, like working in spray-painting booths for days on end. Nobody minded that too much. But then they took over welding, which had always been considered a skilled job. People minded that.

What many of my co-workers didn't seem to notice was that the robotics industry was springing up all round us, right here in Michigan. The auto companies were the customers, so the robotics companies set up shop nearby. The state universities started to offer classes in robotics, just as they had offered classes in automotive technology for decades. So I started to take classes in robotics technology. One class at a time, year-round. Took forever.

As time passed, and more and more assembly line jobs were taken over by robots, I became more valuable to the company even while my co-workers went on strike and demanded that the company pay them for jobs that could be done much cheaper by robots. When I finished my bachelor's degree and was promoted to a full-fledged robotics technician I got a huge raise. Not only that, I got a great deal of personal satisfaction out of making the change. The world doesn't owe me or anybody else a living. You change with the times or drown in them.

Automation doesn't destroy jobs, it creates them. You

may have to go back to school or get some other kind of training to take advantage of the new opportunities, but they're there, right in front of you. That's why I say that some people love my story and some people hate it. Some people have a vested interest in maintaining the status quo and keeping people in their place. They don't like it when a regular person like me proves them wrong. But then there are the other kind of people, the optimists who believe that all people are in charge of their own destiny, and that we must grow and change over time. Those are my kind of people."

PERSONAL QUALITIES

THE ROBOTICS INDUSTRY IS A BROAD and fascinating field that attracts many gifted individuals who want to make their mark. It is also an extremely competitive field. You will be up against serious competition from the day you write your first résumé.

Robotics is a unique field that attracts unique thinkers. It is also propelled in large part by the age-old question: Does life imitate art or does art imitate life? All professionals in the robotics field, whether they admit it or not, want to build the first practical android. They have all been nurtured on a lifetime of science-fiction androids, they have all pondered the ramifications of a world in which robots are both sophisticated and commonplace, and they were all attracted to the robotics industry in some part by their desire to make that sci-fi world come to life. You will be competing against big dreamers with impressive credentials. You must be driven to succeed.

You must also have a very wide and deep creative streak. Most forms of engineering have well-defined end states. That is, when structural engineers are given an assignment to build a bridge between two specific points, that is what they design. They may be able to incorporate a few innovations into the design, but there is no doubt about the goal they are working toward. Robotics engineers are also given assignments with certain specifics, but there will always be endless discussions about how to achieve that result. The question is never about whether to push the technology but how far to push it this time. There is never any doubt in anybody's mind that the next project will always be more amazing than the last. Robotics engineers are always reaching for that next big breakthrough.

Even more important than the next big breakthrough is the next practical application of the last achievement. Incredibly sophisticated technologies are always born a long time before they come into common use. The android technologies pioneered by Honda and other companies at the forefront of robotics engineering are impractically expensive. Such projects are often known as technology demonstrators and see most of their use at trade shows and theme parks, mostly because they are too expensive to be put to any practical use beyond showing themselves off. As an industry insider you will know that there is always a big gap between the latest technology and the latest product for public consumption. As driven and creative as you need to be, you also need to be able to compromise and take pride in small victories. Something you invent today may not make it to the market for many years.

ATTRACTIVE FEATURES

ROBOTICS WOULDN'T BE SUCH A fascinating industry if it didn't have many attractive features. Among those are its unique nature. Robotics engineering is a unique career which embraces several other fields. Most easily described as a synthesis of mechanical and electrical engineering, robotics also touches upon electronics, hardware and software engineering, and industrial design. Advanced robotics research even delves into psychology, because artificial intelligence has a psychology of its own, and because there are many questions regarding human interaction with robots. There is no field quite like robotics, intellectually or from a practical standpoint. If you want to devote your working life to something vitally important but also a little different from everything else, a career in robotics may be for you.

Robotics also holds a unique, and somewhat controversial, place within society. Robots, in simplest terms, were invented and are continuously improved in order to replace human effort. That's their purpose. Factories have replaced humans with robots because they can do repetitive jobs better, faster and at lower costs than humans can. They can also do jobs that would be dangerous to humans and can reach into awkward places that humans can't. The bottom line is that robots cost less over time than human employees do, which reduces the price of the products they create. Robots also increase product quality and uniformity. There is no shortage of debate over whether or not replacing humans for certain types of work is good or bad for society at large, but the fact remains that professionals in the field are front and center in moving this development forward.

The most attractive feature of the robotics industry is its unlimited promise. From battlefields on earth, to fields of unknown elements on planets far away, robots are taking strides that simply would not otherwise be possible. Science fiction writers can get carried away with descriptions of robots completely indistinguishable from humans, or computers with so much artificial intelligence that they run entire planets. Still, those stories are what push robotics engineers to advance the science. This is an industry still in its infancy. You can be a part of helping it to grow up.

UNATTRACTIVE FEATURES

A CAREER IN ROBOTICS IS HARD work. There is no such thing as "easy engineering." Any profession with the word engineering in its title is going to be intellectually challenging. It will require long hours and many headaches. It won't always work out the way you want it to. There will always be people out there who are willing to put in longer hours and endure more headaches than you in order to get to the top. Somebody will always have a better idea. Worse yet, people with ordinary ideas may sometimes do a better job of convincing people with money to finance their projects, relegating your brilliance to the sidelines while someone less worthy steals the show.

Everybody knows that robots are sophisticated visions of the future. Books, television programs and movies featuring robots resonate with wide audiences far beyond serious sci-fi fans. Most people get a kick out of the simple robots in their lives, like kids' toys. But almost nobody knows who creates robots. Even if you are the lead engineer behind the latest and greatest NASA achievement that captures the attention of the entire world, you will never become famous. You'd think that robotics engineers would have a bit of star power. Most people never give robotics a second thought because they assume it's far too complicated – the province of ivory tower geniuses, not regular folk. You may revolutionize the world but your neighbors won't give you much credit for it.

From the dawn of the Industrial Revolution in the early 1700s there have been people who decry automation as the beginning of the end for the human race. Known as Luddites, after a popular movement spearheaded by English textile workers in the early 1800s who took their inspiration from the mythical King Ludd, such people believe that automation destroys humanity by destroying jobs, skills and honest labor. Such beliefs refuse to go away despite centuries of evidence that says automation increases human productivity and broadens opportunities by eliminating dreary jobs. It may be jarring when a robot replaces a human for a job but the truth of the matter is that the cost of the product will go down so more people can buy it, and the person replaced can put time and energy into learning a new skill.

EDUCATION AND TRAINING

THERE ARE MANY EDUCATIONAL paths into a career in robotics. You may have to follow more than one in order to get to where you want to go. Robotics is a unique combination of science and engineering that attracts smart people willing to earn multiple degrees if that's what it takes to set them up to make the next big leap. If you have set your sights on this career be prepared to spend time in school.

Several universities offer specialized bachelor's degree programs in robotics. These programs combine courses in computer science, electrical engineering and mechanical engineering, with additional courses and laboratory time to learn to use those skills in robotics applications. These programs provide a broad overview of the field and tell potential employers that you are specifically interested in a career in robotics, not just a career in some form of engineering or computer science. There are also programs available in robotics technology, which concentrate on the basics of robotics and prepare students to become skilled robotics technicians.

You can also choose to major in one of the fields critical to robotics, and then target your job hunt at employers who will give you an opportunity to apply your skills to the robotics industry. Electrical engineering, mechanical engineering and computer science are all excellent majors for aspiring robotics professionals. When choosing one of these majors you should first think about which direction you would like to take in robotics – robotics engineering or artificial intelligence? You are unlikely to work equally in all areas, and certainly not early in your career. Decide which path interests you the most and pursue it.

A graduate degree should be considered mandatory. You don't need to earn one immediately after you finish college, however. In fact, you should wait until you have worked a few years in the field. You'll have a better idea of what you want to focus on in your further studies. A Master of Science degree will be the most useful target. Not only is robotics a rapidly evolving field in which there is always something new to learn, you will also be competing against academically oriented professionals who will be earning multiple degrees during the course of their careers. If you want to keep up, you will need to keep polishing your credentials – academic credentials and professional credentials. Plan on attending many seminars and conferences.

Should you earn a PhD? That depends upon where your career leads you. If you find that you are happily engaged as a robotics technician or that your employer is providing you with opportunities that you find fulfilling, then probably not. If, on the other hand, you find yourself designing spacecraft for NASA, for example, you may find that a PhD becomes essential. If you do rise to this level you will be among the true leaders in robotics technology. Relatively few industries present such clear-cut opportunities to rise all the way to the top.

No matter what you major in you should not let your undergraduate years go by without completing at least one internship. Most internships are paid, and many come with special advantages like professional development seminars not available to the company's regular employees. You will learn a great deal about robotics simply by working alongside seasoned professionals already in the field, even if your primary function is making coffee and filing reports. You will also make valuable professional contacts that will come in handy when the time comes to find your first full-time job after college. In fact, it is very common for recent graduates to start their careers with the companies where they completed internships. On the other hand, some students change their major after completing an internship concluding that their chosen career wasn't what they thought it would be. This is equally valuable – it's better to make that change while you are still in college than after you've graduated.

EARNINGS

EARNINGS VARY WIDELY IN THE robotics industry but it's safe to say that they are competitive with other fields requiring similar levels of education.

For example, mechanical engineers can expect to earn about $60,000 a year within a few years of graduating from college. That figure can easily double as you rise in seniority and skill. Those with degrees in electrical engineering, computer science and robotics can expect to see similar paychecks. If you get into the business side of things by starting your own company or by inventing products and seeking patents, you could potentially earn millions of dollars.

Many robotics professionals are employed by universities and government agencies, all of which have their own internal pay scales that may not quite match those offered by the private sector. What these institutions do provide, however, is stable employment, generous benefits like pension plans, and, sometimes, enormous research-and-development budgets. If you want to build a Mars rover you will have to work for NASA or for one of a handful of private contractors with the skills to contribute to such a project.

No matter what path you take it behooves you to keep in mind the fact that compensation isn't just about money. In fact, study after study shows that the most important factor in determining job satisfaction isn't money, but reason to believe that you're doing something that matters and that you enjoy. If you are motivated by the promise of pushing boundaries, making discoveries and fundamentally changing the world, you will always be happy to be in the field of robotics.

OPPORTUNITIES

YOU ARE ENTERING THIS FIELD AT AN auspicious time. Job opportunities in robotics are predicted to grow faster than the economy as a whole. This is due to several factors, all of which will benefit you.

First, industrial automation is no longer a curiosity one notch removed from science fiction. Most American factories are largely automated, and this is a trend that will not abate as long as robots remain less expensive than people. The cost of automation has declined to the point that smaller firms that could not or would not invest so much capital in a new venture are now able to do so, creating additional demand for robotics expertise. Most assembly lines and factory floors are custom installations. The same basic robots may be used to assemble cars or kitchen blenders but they have to be configured differently, programmed differently and arranged differently on the factory floor. Many industrial setups actually require new devices to be designed to meet very specific needs. All of these factors lead to demand for skilled robotics professionals.

Artificial intelligence is also entering a new phase, with the stuff of last year's science fiction becoming part of the software bundle in this year's smart phones. Even the definition of what constitutes artificial intelligence has evolved, with practitioners recognizing not only the super-powerful computers that never leave laboratories, but also the relatively simple spinoffs like cloud computing and iPads. As with any industry, the key to growth for robotics and artificial intelligence is in finding a real-world market. Laboratory projects are intellectually stimulating but having millions of customers is what really causes an industry to take off.

The hardest part of your career search may be deciding which specific path you want to take because there are so many to choose from. At this stage you shouldn't worry about it too much. Get your foot in the door and take your time finding your way. You may be surprised at how many opportunities present themselves.

GETTING STARTED

DON'T BE INTIMIDATED WHEN THE time comes to seek out your first real job after college. Polish up your personal marketing materials, get in touch with the contacts you've made while in school and be open to suggestions.

Make sure your résumé is accurate and up to date. This is your single most important marketing tool. If you're not confident in your ability to write a polished résumé, you can seek out help from your school's outplacement office, from a professional résumé writer or from any number of books and software applications. You will need a traditionally formatted version that you can email or print as needed, as well as a simple plain-text version for submitting to employer websites. Resist the urge to exaggerate your accomplishments. In this interconnected age it only takes a few seconds to verify pretty much anything you put on your résumé.

Once you have your résumé in order, you should get in touch with your professional contacts. These could be past employers, professors, and the company with whom you completed your internship. It is very common for companies to offer interns their first real jobs after college. They are known quantities who understand the company. Even if your former employer can't offer you a job, your former managers and colleagues may be able to spread your résumé around or point you in the direction of somebody who is hiring. You may be surprised at how helpful people can be if you just reach out and ask. Everybody wants to see someone new get a good start in a career.

Never lose faith in yourself or your vision for the future. Even if some parts of the future look a bit cloudy from the vantage point of a recent college graduate, charge ahead and take advantage of the opportunities that come your way. Don't worry too much about your first job. Just get one and explore the opportunities it offers. If it's not the perfect fit you can always move along.

ASSOCIATIONS, PERIODICALS, WEBSITES

■ **Abb Robotics**
www.abb.com

■ **Ai Research**
www.a-i.com

■ **Artificial Intelligence Foundation**
www.alice.pandorabots.com

■ **Artificial Intelligence Journal**
www.journals.elsevier.com
/artificial-intelligence

■ **Durabotics**
www.durabotics.com

■ **Edgewater Automation**
www.edgewaterautomation.com

■ **Fanuc Robotics**
www.fanucrobotics.com

■ **Journal of Artificial Intelligence Research**
www.jair.org

■ **Kawasaki Robotics**
www.kawasakirobotics.com

■ **KC Robotics**
www.kcrobotics.com

■ **Kuka Robotics**
www.kuka-robotics.com

■ **Kurzweil Accelerating Intelligence**
www.kurzweilai.net

■ **National Aeronautics and Space Administration**
www.nasa.gov

■ **New Scientist**
www.newscientist.com

■ **PAR Systems**
www.par.com

■ **Popular Mechanics**
www.popularmechanics.com

■ **Popular Science**
www.popsci.com

■ **Progressive Machine and Design**
www.pmdautomation.com

■ **Robot Café**
www.robotcafe.com

■ **Robot Magazine**
www.find.botmag.com

■ **Robotic Industries Association**
www.robotics.org

■ **Robotics Institute**
www.roboticsinstitute.com

■ **Robotics Institute at Carnegie Mellon University**
www.ri.cmu.edu

■ **Robotics Research Lab at the University of Southern California**
www.robotics.usc.edu

■ **Robotics Trends**
www.roboticstrends.com

■ **Science Daily**
www.sciencedaily.com

■ **Science News**
www.sciencenews.org

■ **Scientific American**
www.scientificamerican.com

■ **Stanford Artificial Intelligence Laboratory**
www.ai.stanford.edu

■ **Staubli**
www.staubli.com

■ **Wolf Robotics**
www.wolfrobotics.com

■ **Worcester Polytechnic Institute**
www.wpi.edu

■ **Yaskawa**
www.motoman.com